Quantum world

Atoms

Nucleus Center

Electron Dance

Structure of Atom

Atomic Number

Isotopes

Chemical Bonding

Types of Ch

"Quantum leap"

Molecules and Compounds

Quantum mechanics in our daily lives

Real-time uses Periodic Elements

Quiz

Quantum world

Imagine the quantum world as a magical land that's super, super tiny, where atoms and particles live. Let's explore it in a kid-friendly way

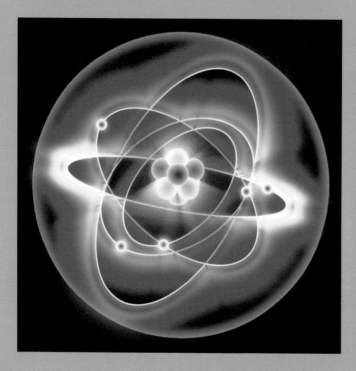

Atoms

Atoms are like the building blocks of everything around us, but super, super tiny.

Nucleus Center:

At the center of each atom, there's a special place called the nucleus. It's like the castle where protons and neutrons hang out.

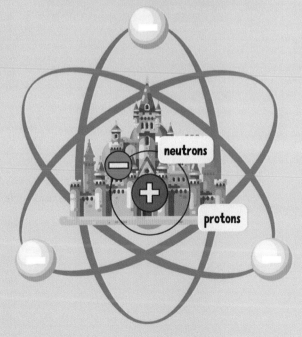

Electron Dance:

Around the nucleus, electrons zoom around like little fairies in different circles called shells.

Superposition Magic:

Sometimes, these fairies can be in more than one place at the same time! It's like magic called "superposition."

Wave-Particle Friends:

Our fairies can be both particles and waves. Imagine them sometimes bouncing like little balls and other times waving like ribbons.

Entanglement Friendship:

Sometimes, fairies become best friends. When this happens, what one fairy does instantly affects the other, no matter how far apart they are!

Quantum Tunnel Adventure:

Our fairies can go on amazing adventures by magically passing through walls — it's called "quantum tunneling."

Colorful Quantum World:

Each fairy has its own magical colors. When they play with light, they create beautiful colors for our eyes to see.

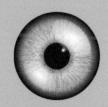

Structure of Atom

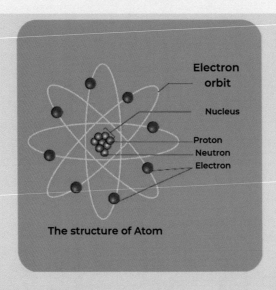

The structure of Atom

Atoms consist of even smaller particles. The three main subatomic particles are:

Protons: Positively charged particles found in the nucleus.

Neutrons: Neutral particles (no charge) also found in the nucleus.

Electrons: Negatively charged particles that orbit the nucleus in electron shells.

Nucleus:

The central part of an atom, called the nucleus, contains protons and neutrons. Protons carry a positive charge, and neutrons have no charge.

Electron Shells:

Electrons move in specific orbits or shells around the nucleus. Each shell can hold a certain number of electrons, and electrons in outer shells are involved in chemical reactions.

Elementary Particles:

Atoms are made up of elementary particles, which include protons, neutrons, and electrons. These particles are, in turn, made up of even smaller entities known as quarks and leptons.

Atomic Number:

The number of protons in the nucleus determines the identity of an element.

This number is called the atomic number.

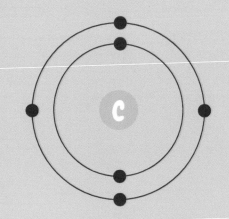

For example, all carbon atoms have six protons.

Isotopes:

Atoms of the same element can have different numbers of neutrons. These variants are called isotopes.

While isotopes of an element have the same number of protons, they may have different atomic masses.

1 Proton
1 Electron
0 Neutrons

1 Proton
1 Electron
1 Neutrons

1 Proton
1 Electron
12 Neutrons

Making Friends - Chemical Bonding

Imagine Friends Holding Hands:

Think of atoms as friends who want to hold hands to feel more stable and happy. This hand-holding represents a chemical bond.

 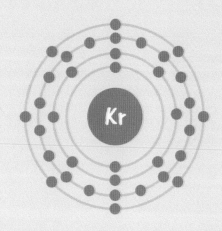

Valence Electrons: Emphasize that electrons in the outermost dance orbit are special. These are called valence electrons and are the ones that like to hold hands.

Types of Chemical Bonds

Ionic Bonds:

Big Donor and Receiver Game:

Imagine one friend has a lot of energy (positive charge) and wants to give it away, and another friend wants to receive that energy.

They swap hands and become best friends. This is like an ionic bond!

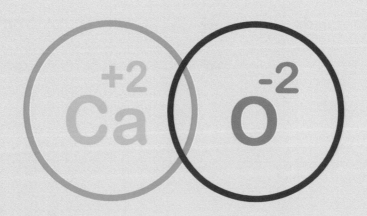

Covalent Bonds:

Sharing is Caring: Other friends decide to share their hands. They link arms and share their energy, like two atoms sharing electrons. This is a covalent bond!

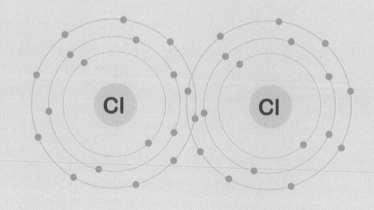

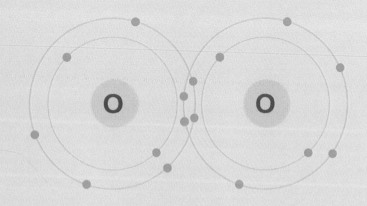

"Quantum leap"

Imagine Dance Floors:

Think of an atom like a dance floor, and electrons are dancers moving around. Each dance floor represents a specific energy level or orbit.

Quantum Leap as a Special Dance Move:

A quantum leap is like a special dance move where an electron jumps from one dance floor to another without passing through the space in between. It's not like a smooth, continuous dance but more like a magical teleportation!

Energy Levels in an Atom:

Atoms have different energy levels, and electrons can only exist on specific dance floors. Each dance floor has a different amount of energy.

Absorption and Emission of Light:

When an electron makes a quantum leap to a higher energy level, it absorbs energy. When it jumps back down to a lower energy level, it releases that energy in the form of light. This is how atoms give off colorful lights.

Example with Fireworks:

Think of a firework. The different colors you see are like the colorful lights emitted when electrons make quantum leaps in atoms. Each color represents a different energy level.

No In-Between Steps:

Unlike regular dancing where you move smoothly from one step to the next, a quantum leap means the electron skips all the steps in between and instantly appears on a new dance floor.

Molecules and Compounds

Molecules: When atoms bond, they become a molecule. Imagine a molecule as a group of friends holding hands and dancing together.

Compounds: Groups of different types of atoms (elements) holding hands form compounds. For example, water is a compound made of hydrogen and oxygen atoms holding hands.

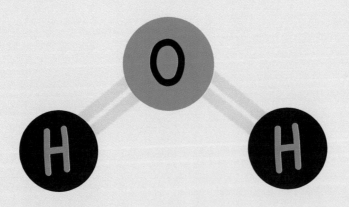

NaCl (Salt): Sodium and chlorine atoms holding hands in a special way (ionic bond).

Quantum mechanics is utilized in our daily lives

1.Semiconductor Devices (Computers and Smartphones):

Simple Explanation: Imagine tiny electronic helpers inside your computer or smartphone. These helpers, called semiconductors, help control the flow of information and make your devices work.

2. MRI (Magnetic Resonance Imaging):

Simple Explanation: An MRI is like a magic camera that takes pictures inside your body using strong magnets. These magnets make the atoms in your body behave in a special way, helping doctors see detailed images.

LED (Light-Emitting Diodes):

Simple Explanation: LEDs are like tiny magic light bulbs. When you give them a little energy, they glow with bright colors. They are used in toys, gadgets, and even traffic lights.

4. Nuclear Energy:

Simple Explanation: Imagine a special type of energy hidden inside the center of atoms. We can use this energy, called nuclear energy, to generate electricity. It's like unlocking a powerful energy source from the tiniest building blocks of matter.

5. Laser:

Simple Explanation: A laser is like a focused flashlight that shoots out very concentrated light. It's used for many things, like pointing in presentations, reading barcodes, or even in surgery to cut things very precisely.

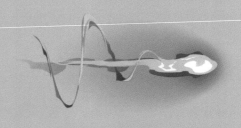

6. Quantum Dots in TV Display:

Simple Explanation: Think of your TV like a magic window. Quantum dots are like tiny colorful fairies that create the bright and vibrant colors you see on the screen. They make your favorite shows and games look super colorful.

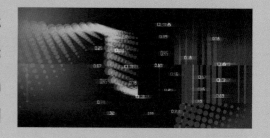

7. GPS Technology:

Simple Explanation: GPS is like a superhero map in your phone. It talks to satellites high above the Earth, asking for signals. By calculating how long it takes for the signals to reach your phone, it figures out exactly where you are. It's your guide on the road!

Periodic Table

The Periodic Table of the Elements

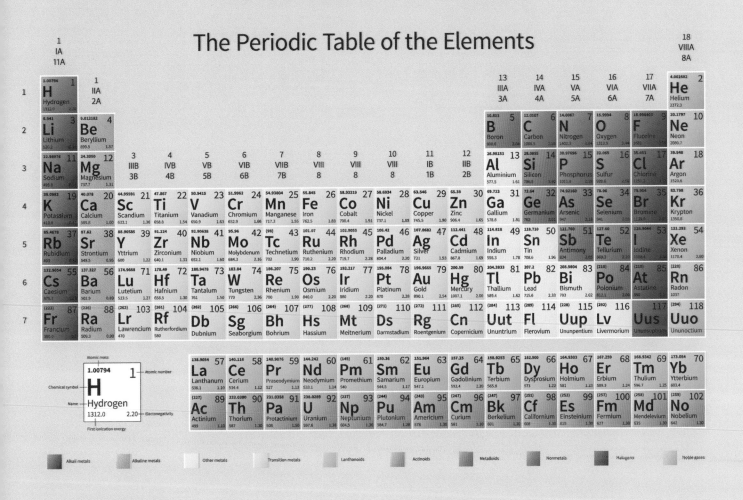

Real-time uses Periodic Elements

1.Hydrogen (H): hydrogen is the primary element present in the Sun. It makes up about 75% of the Sun's mass, while helium accounts for roughly 24%. These two elements dominate the composition of the Sun

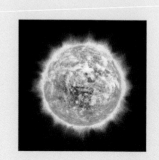

2. Helium (He): Commonly used in balloons, as a cooling medium in medical MRI scanners, and in various industrial applications.

3.Lithium (Li): Used in rechargeable batteries, particularly in mobile phones, laptops, and electric vehicles.

4.Beryllium (Be): Found in various alloys used in aerospace components, X-ray windows, and electronic connectors.

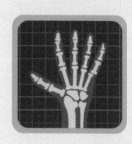

H 1 $1s^1$	He 2 $1s^2$	Li 3 $1s^22s^1$	Be 4 $1s^22s^2$

B 5 $1s^22s^22p^1$	C 6 $1s^22s^22p^2$	Nitrogen N 14.007 7	O 8 $1s^22s^22p^4$

F 9 $1s^22s^22p^5$	10 Ne 20.18 Neon	Sodium Na 22.99 11	12 24.305 Mg Magnesium

5. Boron (B): Boron is used in making sports equipment such as tennis rackets, golf clubs, bicycles, archery equipment, and skis/snowboards due to its strength, stiffness, and lightweight properties, enhancing performance in these activities.

6.Carbon (C): Carbon is indeed considered the basic life molecule because of its unique ability to form a wide variety of complex molecules essential for life.

7.Nitrogen (N): It is used to make fertilisers, nitric acid, nylon, dyes and explosives.

8. Oxygen (O): oxygen is a vital component of air. It makes up approximately 21% of the Earth's atmosphere by volume. Oxygen is essential for the respiration of many organisms, including humans, animals, and most aerobic microorganisms.

9. Fluorine (F): Used in the production of various chemicals, including fluorocarbons, and in dental care products.

10. **Neon (Ne):** Neon is mainly used in lighting, particularly in neon signs. It's also used in indicator lamps, high-voltage indicators, and cryogenic applications.

11. **Sodium (Na):** Used in the production of sodium chloride (table salt) and in various industrial processes, such as detergents.

12. **Magnesium (Mg):** Magnesium is a central component of chlorophyll, the pigment responsible for giving plants their green color. It plays a crucial role in photosynthesis, where it helps capture sunlight and convert it into chemical energy that plants use to produce glucose and oxygen.

13.Aluminum (Al): Aluminum is extensively used in airplanes due to its combination of lightweight properties and strength. It is utilized in the construction of the aircraft's body, wings, and various structural components. This helps reduce overall weight, improving fuel efficiency and performance while maintaining structural integrity and safety standards..

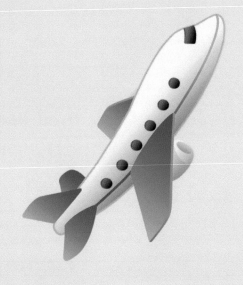

14. Silicon (Si): Essential in the electronics industry for making semiconductors and solar cells.

15. Phosphorus (P): Used in fertilizers, detergents, and the production of matches and fireworks.

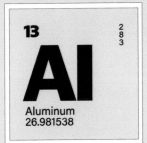

13
2 8 3
Al
Aluminum
26.981538

14
2 8 4
Si
Silicon
28.0855

15
2 8 5
P
Phosphorus
30.973761

16
2 8 6
S
Sulfur
32.066

17
2 8 7
Cl
Chlorine
35.453

Argon
Ar
39.948
18

19
2 8 8 1
K
Potassium
39.0983

20
2 8 8 2
Ca
Calcium
40.078

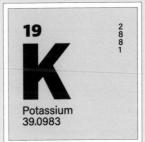

FERTILIZER

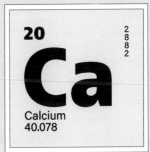

16. Sulfur (S):   Sulfur is present in eggs as sulfur-containing compounds, primarily in the protein-rich egg whites. It contributes to the characteristic smell of cooked eggs and plays a role in protein structure and function.

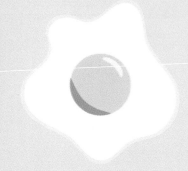

17. Chlorine (Cl): Chlorine is commonly used in swimming pools as a disinfectant to kill harmful bacteria and microorganisms that can thrive in water. It helps maintain the water's cleanliness and safety for swimmers by effectively sanitizing the pool and preventing the spread of waterborne illnesses.

18. Argon (Ar): Argon is used in incandescent light bulbs to fill the bulb and displace oxygen. This prevents the filament from oxidizing and burning out too quickly. Argon also helps to maintain a stable environment inside the bulb, prolonging the life of the filament and improving the efficiency of the bulb.

19. Potassium (K): Important in agriculture as a component of fertilizers, and crucial for various physiological functions in living organisms.

20. Calcium (Ca): Essential for the formation of bones and teeth, used in construction materials like cement, and in the food industry.

21. Scandium (Sc): Scandium is used in bicycle frames to create high-performance alloys, enhancing strength and reducing weight for improved cycling performance.

22. Titanium (Ti): Known for its strength-to-weight ratio, it is used in aerospace, medical implants, and sporting goods.

21
Sc
Scandium
44.955910

22
Ti
$3d^2 4s^2$

23
V
$3d^3 4s^2$

24
Cr

25
Mn
54.94
Manganese

26
Fe
55.84
Iron

27
Co
58.93
Cobalt

28
Ni
58.69
Nickel

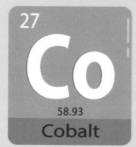

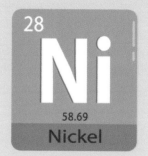

23. Vanadium (V): Vanadium is sometimes used in the production of high-strength steel, including spring steel

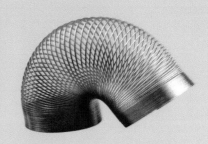

24. Chromium (Cr): Chromium is a key element in stainless steel, contributing to its corrosion resistance and durability. By forming a thin oxide layer on the surface of the steel, chromium helps prevent rust and staining, making stainless steel suitable for a wide range of applications, including kitchen appliances, utensils, and architectural structures.

25. Manganese (Mn): Manganese is used in the construction of earthmoving equipment, such as excavators and bulldozers, as an alloying element in steel. It enhances the strength, hardness, and wear resistance of the steel components, improving the durability and performance of the machinery in rugged environments.

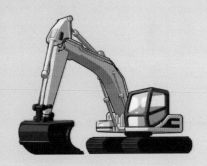

26. Iron (Fe): Essential in the production of steel, used in construction, transportation, and various industrial applications.

27. Cobalt (Co): Cobalt is used in aerospace for engine components, in electronics for lithium-ion batteries, in medical implants, as catalysts in the chemical industry, and in magnets for various applications

28. Nickel (Ni): Used in the production of stainless steel, as well as in various alloys and batteries.

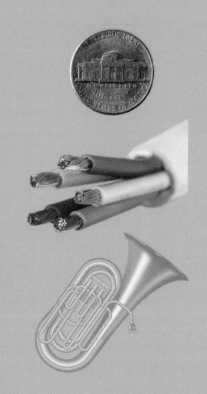

29. Copper (Cu): Widely used in electrical wiring, plumbing, and various electronic devices.

30. Zinc (Zn): paints, rubber, cosmetics, pharmaceuticals, plastics, inks, soaps, tuba, batteries, textiles and electrical equipment.

29
2 8 1
Cu
Copper
63.546

30
2 8 18 2
Zn
Zinc
65.409

31
Ga
69.72
Gallium

32
Ge
72.64
Germanium

33
As
74.92
Arsenic

34
2 8 18 6
Se
Selenium
78.96

35
Br
79.90
Bromine

36
Kr
83.80
Krypton

37
Rb
85.47

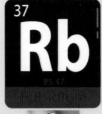

38
Sr
87.62
Strontium

39
Y
88.91
Yttrium

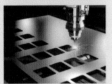

40
Zr
91.22
Zirconium

41
Nb
92.91
Niobium

42
Mo
Molybdenum

43
Tc
[98]
Technetium

31. Gallium (Ga): Gallium is able to turn electricity into light, so it's also used to make LEDs. It can also be used to make thermometers and mirrors.

32. Germanium (Ge): Used in the semiconductor industry and in optical devices like infrared detectors.

33. Arsenic (As): Wallpaper, beer, wine, sweets, wrapping paper, painted toys, sheep dip, insecticides, clothing, dead bodies, Arsenic is a nasty poison that will kill almost any living organism.

34. Selenium (Se): Selenium is used in photocopiers, glassmaking, electronics, chemical processes, and as a dietary supplement due to its antioxidant properties.

35. Bromine (Br): Bromine has a large variety of uses including in agricultural chemicals, insecticides, dyes, pharmaceuticals, flame-retardants, furniture foam, gasoline, plastic casings for electronics, and film photography,

36. Krypton (Kr): Used in certain types of lighting, such as high-powered lasers and photographic flashes.

37. Rubidium (Rb):
Rubidium is used in high-precision atomic clocks for global navigation systems like GPS, ensuring accurate timekeeping essential for determining precise locations.

38. Strontium (Sr): Used in the production of fireworks (strontium compounds provide red colors) and in some medical imaging procedures.

39. Yttrium (Y): Used in LED lighting, certain types of lasers, and in the production of superconductors.

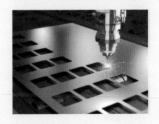

40. Zirconium (Zr): Zirconium is used in chemical pipelines for its corrosion resistance, particularly in environments with highly corrosive substances.

41. Niobium (Nb): Niobium is used in the construction of some parts of high-speed trains, particularly in the manufacture of superconducting magnets for magnetic levitation (maglev) trains.

42. Molybdenum (Mo): Molybdenum is commonly used in the production of cutting tools, such as drills, milling cutters, and taps, due to its excellent strength, hardness, and resistance to high temperatures.

43. Technetium (Tc): Primarily used in medical imaging as a radioactive tracer.

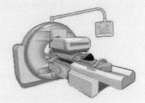

44. Ruthenium (Ru): Used in some types of electrical switches, in the electronics industry, and as a catalyst.

45. Rhodium (Rh): Valued in the automotive industry for catalytic converters, search light reflection and in the production of certain types of jewelry.

44 Ru 101.07 Ruthenium
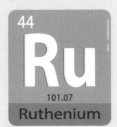

45 Rh 102.91 Rhodium
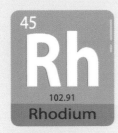

46 Pd 106.42 Palladium

47 Ag 107.87 Silver

48 Cd Cadmium 112.414

49 In Indium 114.818
2 8 18 18 3

50 Sn Tin 118.710
2 8 18 18 4

51 Sb Antimony 121.76
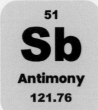

52 Te Tellurium 127.6

53 I 126.90 Iodine

54 Xe 131.294

55 Cs Cesium 132.905

56 Ba Barium

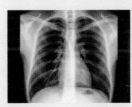

57 La 138.91 Lanthanum

58 Ce Cerium 140.116
2 8 18 19 2
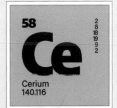

46. Palladium (Pd):

Palladium is used in pollution control devices, particularly catalytic converters in vehicles. In these converters, palladium acts as a catalyst to facilitate the conversion of harmful pollutants, such as carbon monoxide, hydrocarbons, and nitrogen oxides, into less harmful substances like carbon dioxide, water, and nitrogen. This helps reduce emissions from vehicles, contributing to cleaner air and reduced environmental pollution.

47. Silver (Ag): Widely used in jewelry, photography, electronics, and as a catalyst in some chemical processes.

48. Cadmium (Cd): Cadmium was once used in paint pigments for vibrant colors, but due to toxicity, its use has been restricted or eliminated in many countries to reduce health risks from ingestion or inhalation.

49. Indium (In): Used in the production of electronic devices like touchscreens and solar cells.

50. Tin (Sn): Used in the production of solder, as well as in the manufacturing of cans and certain alloy

51. Antimony (Sb): Used in car batteries

52. Tellurium (Te): Used in Thermoelectric cooler

53. Iodine (I): Used in medical applications, such as antiseptics and contrast agents, and in the production of certain chemicals.

54. Xenon (Xe): Used in certain types of lighting, including high-intensity lamps and electronic flash tubes.

55. Cesium (Cs): Used in atomic clocks, drilling fluids in the oil industry, and in certain medical applications.

56. Barium (Ba): Used in X ray diagnosis

57 -71 rare earth metals

57. Lanthanum (La): Used in the production of catalysts, as well as in the manufacturing of camera lenses and high-performance alloys.

58. Cerium (Ce): Cerium is one of the rare chemicals, that can be found in houses in equipment such as color televisions, fluorescent lamps, energy-saving lamps and glasses

59. Praseodymium (Pr): Used in certain types of magnets, especially in wind turbines and electric vehicles.

60. Neodymium (Nd): used in hard disc drives, mobile phones, video and audio systems of television

59	2 8 18 21 8 2
Pr	
Praseodymium 140.90765	

60	2 8 18 22 8 2
Nd	
Neodymium 144.24	

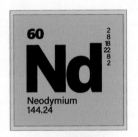

61
Pm
Promethium (145)

62
Sm
150.36 Samarium

63	2 18 25 8 2
Eu	
Europium 151.964	

64
Gd
157.25 Gadolinium

65
Tb
158.93 Terbium

66
Dy
162.50 Dysprosium

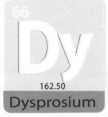

67
Ho
Holmium

68	2 8 18 30 8 2
Er	
Erbium 167.259	

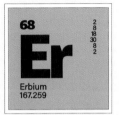

69
Tm
168.93 Thulium

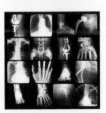

70	173.05
Yb	
Ytterbium	

71
Lu
174.97 Lutetium

72
Hf
178.49 Hafnium

73
Ta
Tantalum 180.948

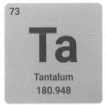

61. Promethium (Pm): Primarily used in nuclear batteries and some luminous paint applications.

62. Samarium (Sm): The most important application of Samarium is in Samarium-Cobalt magnets which have a very high permanent magnetisation. These magnets are used in headphones, small motors and musical instruments like guitars.

63. Europium (Eu): Critical for red phosphors in LED lighting and in certain types of security features in banknotes.

64. Gadolinium (Gd): It is also used in alloys for making magnets, electronic components and data storage disks

65. Terbium (Tb): Essential for green phosphors in LED lighting and in the production of certain electronic devices.

66. Dysprosium (Dy): Used in the production of high-strength magnets, particularly in electric vehicles and wind turbines.

67. Holmium (Ho): Used in some medical lasers and as a component in certain types of magnets.

68. Erbium (Er): Used in fiber optic communications and as a dopant in lasers.

69.Thulium (Tm): Used in some medical lasers and in portable X-ray devices.

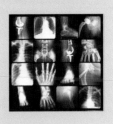

70. Ytterbium (Yb): Used in certain types of lasers, particularly in industrial and medical applications.

71. Lutetium (Lu): Used in certain medical imaging devices and as a catalyst in certain chemical reactions.

72. Hafnium (Hf): Used in the nuclear submarines

73. Tantalum (Ta): Used in the mobile phones

74. Tungsten (W): Known for its high melting point, it is used in the production of lightbulb filaments, electrical contacts, and in the aerospace industry.

75. Rhenium (Re): Used in the rocket engines

74 **W** Tungsten 183.84	75 **Re** Rhenium 186.207 2 8 18 32 13 2	76 **Os** Osmium 190.23	77 **Ir** 192.217	78 **Pt** 195.08 Platinum

79 **Au** 196.97 Gold	80 **Hg** 200.59 Mercury	81 **Tl** Thallium 204.3833 2 8 18 32 18 3	82 **Pb** Lead 207.2	83 **Bi** Bismuth 208.98
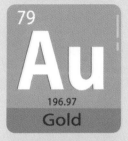			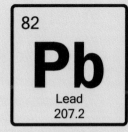	

84 **Po** Polonium 208.982	85 **At** Astatine 209.987	86 **Rn** Radon (222)	87 **Fr** Francium (223)	88 **Ra** Radium 226.025

PLUMBING SERVICES

76. Osmium (Os): Used in pen points

77. Iridium (Ir): Valued for its corrosion resistance, used in electrical contacts, spark plugs, and certain types of medical devices.

78. Platinum (Pt): Widely used in jewelry, catalytic converters, and in the production of certain electronic components.

79. Gold (Au): Highly valued for jewelry, as a store of value, and in the electronics industry for its conductivity.

80. Mercury (Hg): Historically used in thermometers and barometers, but its use has decreased due to environmental concerns.

81. Thallium (Tl): Historically used in some types of rat poisons, but its use has been restricted due to toxicity. It has limited practical applications.

82. Lead (Pb): Used in batteries, radiation shielding, and historically in plumbing and paint (though its use in these areas has decreased due to health concerns).

83. Bismuth (Bi): Used in some pharmaceuticals, cosmetics, and as an alloying agent in low-melting-point alloys.

84. Polonium (Po): Used in radio active brushes

85. Astatine (At): Highly radioactive and scarce; it has no significant practical uses and is primarily studied for research purposes.

86. Radon (Rn): Radioactive gas, and its isotopes are used in some medical treatments and in certain types of research.

87. Francium (Fr): Extremely rare and highly radioactive; it has no practical uses and is mainly studied for research purposes.

88. Radium (Ra): Radium's main practical use has been in medicine, producing radon gas from radium chloride to be used in radiotherapy for cancer.

89. Actinium (Ac): Radioactive and mainly studied for research purposes; it has no significant practical applications.

90. Thorium (Th): Primarily used in nuclear reactors for fuel, and also in certain types of research.

91. Protactinium (Pa): Primarily of interest in nuclear research and doesn't have significant practical applications outside of that field.

92. Uranium (U): Mainly used as fuel in nuclear reactors for electricity generation and in the production of nuclear weapons.

93. Neptunium (Np): Primarily used in nuclear research and has no significant practical applications.

94. Plutonium (Pu): Used as fuel in nuclear reactors, and certain isotopes have been used in the production of nuclear weapons.

95. Americium (Am): Found in some types of smoke detectors and has potential uses in certain types of nuclear batteries.

96. Curium (Cm): Primarily used in research and has no practical applications outside of that context.

97. Berkelium (Bk): Used in research, particularly in the synthesis of heavy elements, and doesn't have practical applications.

98. Californium (Cf): Primarily used in research and doesn't have significant practical applications.

99. Einsteinium (Es): Used in research and doesn't have practical applications outside of scientific investigations.

100. Fermium (Fm): Primarily used in research and doesn't have practical applications beyond that.

101. Mendelevium (Md)	110. Darmstadtium (Ds)
102. Nobelium (No)	111. Roentgenium (Rg)
103. Lawrencium (Lr)	112. Copernicium (Cn)
104. Rutherfordium (Rf)	113. Nihonium (Nh)
105. Dubnium (Db)	114. Flerovium (Fl)
106. Seaborgium (Sg)	115. Moscovium (Mc)
107. Bohrium (Bh)	116. Livermorium (Lv)
108. Hassium (Hs)	117. Tennessine (Ts)
109. Meitnerium (Mt)	118. Oganesson (Og)

Elements 100 to 118including those in the transactinide series, are highly synthetic and typically exist for very short durations in laboratories. They are not found in nature in appreciable amounts, and their practical applications are limited to scientific research.

1.What are the tiny building blocks of everything around us called?

a) Blocks

b) Atoms

c) Bricks

d) Pieces

2. What do you get when atoms join together and become a team?

a) Elements

b) Particles

c) Molecules

d) Blocks

3.What are the tiny dancers that move around an atom called?

a) Fairies

b) Electrons

c) Elves

d) Pixies

4. What are the friendly helpers inside an atom's center?

a) Protons

b) Neutrons

c) Electrons

d) Quarks

5. What is the term for the special friendship handshake between atoms?

a) Atomic hug
b) Molecule bond
c) Chemical bond
d) Electron shake

6. What are the tiny electronic helpers inside computers and smartphones called?

a) Microchips
b) Semiconductors
c) Electromagnets
d) Gizmos

7. What do we call the tiny magic light bulbs that light up with bright colors?

a) Glow bulbs

b) Light bulbs

c) LED

d) Spark bulbs

8. What is like a superhero map in your phone that talks to satellites?

a) MapQuest

b) GPS

c) Satellite Guide

d) Star Map

9. Where is the special kind of energy hidden inside the center of atoms?

a) Atomic Power
b) Nuclear Energy
c) Core Energy
d) Elemental Force

1. b) Atoms
2. c) Molecules
3. b) Electrons
4. b) Neutrons
5. c) Chemical bond
6. b) Semiconductors
7. c) LED
8. b) GPS
9. b) Nuclear Energy

15 Quiz questions on periodic table

1. Hydrogen (H) is the _____ element in the periodic table.

2. The chemical symbol for Gold is _____.

3. Carbon (C) is the building block of _____.

4. The symbol for Iron is _____.

5. Oxygen (O) is a gas essential for _____.

6. Silver (Ag) is commonly used in the production of _____.

7. Aluminum (Al) is known for being _____, making it widely used in packaging.

8.The chemical symbol for Lead is ____.

9. Copper (Cu) is widely used in electrical ____.

10. Helium (He) is often used to fill ____.

11. The symbol for **Mercury** is ____.

12. Nitrogen (N) makes up a significant portion of Earth's ____.

13. Chlorine (Cl) is commonly used in ____.

14. The chemical symbol for Sodium is ____.

15. Zinc (Zn) is often used in the production of ____.

1. First
2. Au
3. Life
4. Fe
5. Respiration
6. Jewelry
7. Lightweight
8. Pb
9. Wiring
10. Balloons
11. Hg
12. Atmosphere
13. Water purification
14. Na
15. Alloys

CERTIFICATE

OF ACHIEVEMENT

Presented to :

_____

completing your exploration of quantum physics
and the periodic table..

Rector

Esmart_chubs

Printed in Great Britain
by Amazon

47281179R00034